FROM EDDYSTONE TO THE STARS

The Astronomy of John Smeaton

(1724-92)

David Sellers

2024

Magavelda Press
38 Gledhow Wood Avenue
Leeds LS8 1NY
UK

Copyright © David Sellers 2024

First published by MagaVelda Press 2024

(2nd impression)

ISBN 978 0954101329

Smeaton

the 'Natural Philosopher'

I n James Ferguson's celebrated book, *Astronomy Explained Upon Sir Isaac Newton's Principles* (1778), there is a table for the *Equation of Time*. Ferguson says that the table is 'by Mr Smeaton'.[1]

This is John Smeaton (1724-92), who is justly famed for the design and construction of the third Eddystone Lighthouse. Many of his great engineering works— mills, harbours, bridges and waterways—are still in use today. Smeaton was the archetypal civil engineer, as opposed to the previous mainly military engineers.

There are not many monuments or plaques in memory of Smeaton, but those that do exist all make reference exclusively to his works of civil engineering—especially the third Eddystone Lighthouse—and maybe to his founding of the 'Society of Civil Engineers' (the forerunner of the modern Institution of Civil Engineers)

Fig.1 John Smeaton, c.1788

This even applies to the most unlikely of memorials, such as the church plaque (fig.3) and vestments at the Parish church of St Mary at Whitkirk, which both depict the famous lighthouse.

The *Equation of Time* table (see fig.2), which John Smeaton provided for Ferguson, gives figures for how much a clock will be faster or slower than the Sun

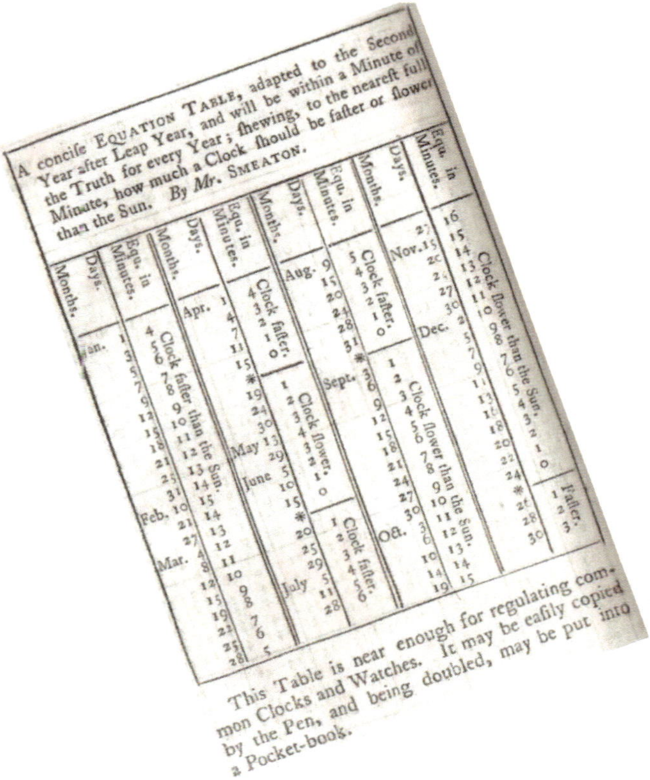

A conciſe EQUATION TABLE, adapted to the Second Year after Leap Year, and will be within a Minute of the Truth for every Year; ſhewing, to the neareſt full Minute, how much a Clock ſhould be faſter or ſlower than the Sun. By Mr. SMEATON.

Months.	Days.	Equ. in Minutes.	Months.	Days.	Equ. in Minutes.	Months.	Days.	Equ. in Minutes.	Months.	Days.	Equ. in Minutes.
Jan.	1	4 Clock faſter than the Sun.	Apr.	1	4 Clock faſter.	Aug.	9	5 Clock faſter.		27	16 Clock ſlower than the Sun.
	3	5		4	3		15	4	Nov.	15	15
	5	6		7	2		20	3		20	14
	7	7		11	1		24	2		24	13
	9	8		15	0		28	1		27	12
	12	9		*			31	0		30	11
	15	10		19	1 Clock ſlower.		*		Dec.	2	10
	18	11		24	2	Sept.	3	1 Clock ſlower than the Sun.		5	9
	21	12		30	3		6	2		7	8
	25	13	May	13	4		9	3		9	7
	31	14		29	3		12	4		11	6
Feb.	10	15	June	5	2		15	5		13	5
	21	14		10	1		18	6		16	4
	27	13		15	0		21	7		18	3
Mar.	4	12		*			24	8		20	2
	8	11		20	1 Clock faſter.		27	9		22	1
	12	10		25	2		30	10		24	0
	15	9		29	3	Oct.	3	11		*	
	19	8	July	5	4		6	12		26	1 Faſter.
	22	7		11	5		10	13		28	2
	25	6		28	6		14	14		30	3
	28	5					19	15			

This Table is near enough for regulating common Clocks and Watches. It may be eaſily copied by the Pen, and being doubled, may be put into a Pocket-book.

Fig.2 Smeaton's Equation of Time table from Ferguson's Astronomy Explained

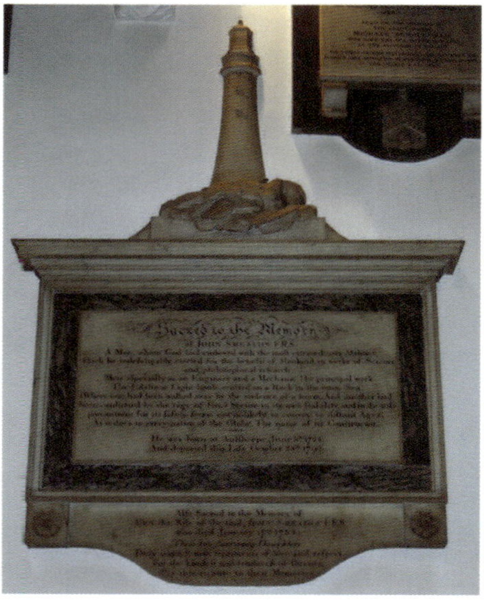

Fig.3 Smeaton's memorial plaque at St Mary's Church, Whitkirk, Leeds

at different times of the year. Public sundials are often accompanied by a graph or table showing these corrections to bring sundial time into line with mean time (i.e. clock time) throughout the year, though few passers-by these days appreciate the difference.

The discrepancy between clock time and sundial (or solar) time arises *firstly* from the fact that the Earth is travelling round the sun in an ellipse at varying speed (faster in the winter when we are nearer to the sun), and *secondly* because the spin axis of the Earth is tilted (relative to the plane of orbit) at an angle of about 23½° from the vertical (see fig.4). Calculation of the required correction (or 'equation') is not trivial.

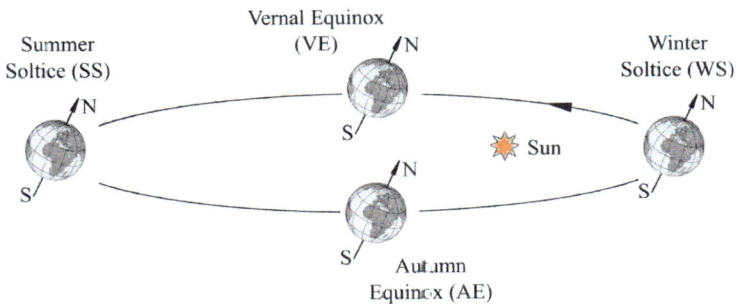

Fig.4 The Earth orbits the Sun on an elliptical path with its spin axis inclined relative to the plane of the ellipse

So, how did it come to pass that an arcane astronomical table should be provided by a civil engineer? This booklet aims to provide an explanation.

Although Smeaton's later professional life was dominated by the civil engineering works for which he is still honoured, his election to the Royal Society at the age of 28 preceded any of these works and was the result of his earlier role as a maker of 'philosophical', or scientific, instruments. Throughout his life his leisure hours were dominated by an intense interest in 'natural philosophy'—as science was then known—and especially in astronomical matters. Undeniably, such matters were the only things allowed to occasionally impinge upon his engineering work. His daughter, Mary, later recounted that

> his time was governed by a method as invariable as inviolable: for professional studies were never broken in upon by any one; and these (with the exception of stated astronomical observations) wholly ingrossed the forenoon.[2]

Indeed, the Preface to the 1837 edition of Smeaton's 'Reports' (*Some Account of the Life, Character and Works of Mr John Smeaton, FRS*), relates that

> Astronomy was one of Mr.Smeaton's most favourite studies; and he contrived and made several astronomical instruments for himself and friends. After fitting up an observatory at his house at Austhorpe, he devoted much of his time to it when he was there: even in preference to

Fig.5 Smeaton's home: Austhorpe Lodge (sketched by Percival Skelton, after an original drawing by T. Sutcliffe)

Fig.6 Remains of Smeaton's workshop shortly before demolition in the early 1930s

public business, much of which he declined for the purpose of applying his attentions to private study, particularly to the subject of astronomy.[3]

John Smeaton was born at Austhorpe Lodge, in Whitkirk, East Leeds, on 8th June 1724 and lived there for much of his life (figs.5 & 6). Unfortunately, the Lodge was demolished in the 1930s to make way for a housing estate. Here we would have seen the the tower that he built, on top of which he made his astronomical observations. One floor lower was the design office for his professional work and at the base of the tower was the workshop, where many of his experiments were conducted and his instruments were fabricated.

Smeaton's father had a legal practice in Leeds, but from an early age John seemed to be drawn more to the profession of his grandfather, who had been a watchmaker of some renown in York. Already, by the time he left Leeds Grammar School at the age of 16 he had his own workshop. A distant cousin (John Holmes), visiting John two years later, reported that

> in the year 1742, I spent a month at his father's house, and being intended myself for mechanical employment, a few years younger than he was, I could not but view his works with astonishment; he forged his iron and steel, and melted his metal; he had tools of every sort … He had made a Lathe, by which he cut a perpetual screw in brass, a thing little known at that day, and which I believe was the invention of Mr Henry Hindley, of York …. Mr Hindley was a man of the most communicative disposition, a great lover of Mechanics, and of the most fertile genius. Mr Smeaton soon became acquainted with him, and they spent many a night at Mr Hindley's house till day light, conversing on those subjects.[4]

The prominence given to the perpetual screw, mentioned in this account, perhaps owes something to Smeaton's own estimation of its importance. This screw was essentially a precision worm gear for the careful adjustment of setting circles and arcs on astronomical instruments so as to achieve accurate measurements. The screw had an 'hour-glass'—rather than cylindrical—shape, which was calculated to reduce any play (or 'what workmen term drunkenness') in the gearing. Its manufacture required a sophisticated dividing engine and Smeaton rightly saw such devices as being crucial for the advancement of astronomy.[5]

Smeaton first met the screw's inventor, Henry

Hindley (1701-71), 23 years his senior, in 1741. The older man had a successful clock-making business in York, but had also engaged in making telescopes, a steam engine, a pyrometer, and many other mechanical devices. He had very recently completed a sophisticated 'dividing engine', for graduating arcs and circles, at his workshop near Lendal Bridge in York.

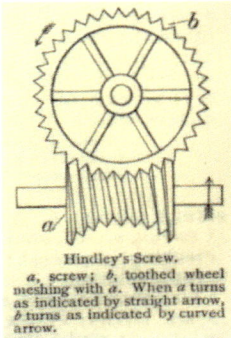

Hindley's Screw.
a, screw; *b*, toothed wheel meshing with *a*. When *a* turns as indicated by straight arrow, *b* turns as indicated by curved arrow.

Fig.7 Hindley's 'perpetual screw' (from The Century Dictionary of WD Whitney (1895), p.5420)

Remarkably open and generous as he was with the young Smeaton, Hindley nevertheless kept some of the manufacturing process for his 'hour-glass' screw a 'trade secret'.

The Instrument Maker
and Experimenter

Not long after his first acquaintance with Hindley, the young Smeaton was sent to London to commence a legal education—perhaps by parents who were becoming alarmed at their son's bent for mechanical matters rather than more gentlemanly occupations. Within two years he returned, persuaded that legal circles were not for him. Whilst in London, however, he had met and formed crucial friendships with Benjamin Wilson and other members of the Royal Society—the first and foremost scientific academy in the country. He continued a correspondence with these friends, which frequently dealt with astronomy. Soon he resolved to make a profession out of building scientific instruments. This was to be a lifelong passion. By the age of 23 he had built his own telescope and had become, in his own estimation 'quite an artist' at grinding and polishing lenses.[6]

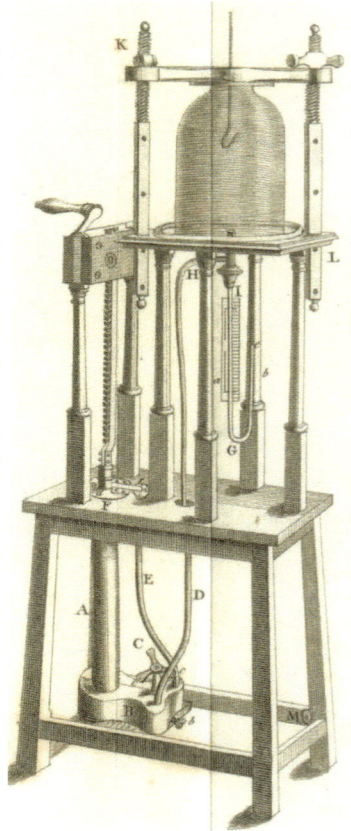

Fig.8 Smeaton's vacuum pump, as depicted in his 1752 paper to the Royal Society

Vacuum Pump

In 1748, aged 26, with parental blessing, he departed

once more for London to start his instrument building career. The first product of this endeavour was a vacuum pump that ended up being used, by Wilson and others, in electrical discharge experiments. In a later book on the properties of light, Wilson described the use to which he had put the pump:

> ...from some experiments I made many years ago, in company with Mr. Smeaton, and with his excellent air pump, which, when it was in very good order, rarefied the air above two thousand times, we observed, in general, that very small differences of air occasioned very material differences in the luminous effects produced by the electric fluid; insomuch, that when all the air was taken out of the receiver, which this pump, at that time, was capable of extracting, no electric light was perceptible in the dark: upon letting in a little air by a stop-cock, a faint electric light was visible, and the letting in of more air increased the light considerably. But this light began to decrease on the letting in of more air, till, at last, on letting in greater quantities, it entirely vanished.[7]

Many years later the French astronomer, Jérôme Lalande (1732-1807), during a trip to England, alive to the significance of the pump, mentioned it in his diary:

> Saturday 27th May 1763.

> We had lunch ... and we went to the house of the Queen. ... the King gave an order that his pneumatic machine be brought at 10:00 to have us see it. ... This machine is designed by Mr Smeaton who had come from Yorkshire, having much talent for machines ... With it one can rarefy two thousand times compared with ordinary machines, which can only go to three hundred times.[8]

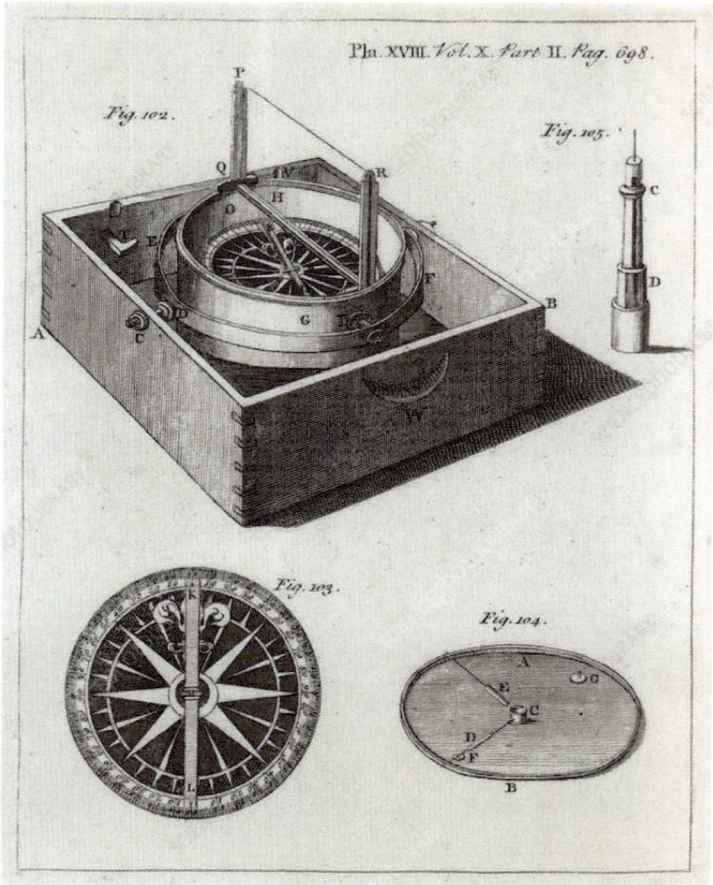

Pla. XVIII. *Vol. X. Part II. Pag. 698.*

Fig. 9 The improved Marine Compass (Philosophical Transactions of the Royal Society, v.46 (1750), 513-17)

Marine Compass

One of the next fruits of Smeaton's industry was a marine compass (developed in collaboration with Dr Gowin Knight). He submitted a paper on this to the Royal Society in July 1750.[9] It is interesting to note that at the head of this paper he still styles himself 'Philosophical [i.e. Scientific] Instrument-maker'.

Other high quality instruments followed – including a telescope for the clock-maker, John Ellicott, and a precision lathe for William Matthews. Around this time he also began his pioneering experiments on the power of water and wind. All in all, it was clear that, as well as being a formidable instrument builder, Smeaton was a first class scientist. In December 1752, he was proposed by Lord Charles Cavendish for membership of the Royal Society and in March 1753, with the support of the telescope-maker, James Short, and others, he was duly elected.

Transit of Mercury, 1753

One of the hallmarks of Smeaton's commitment to instrument building was his pursuit of measurement accuracy and precision. It was around this time, on the occasion of a transit of the planet Mercury across the face of the Sun, that we start to see the public expression of Smeaton's application of this to astronomical observations.

According to a report of 31 May 1753, addressed to Sylvanus Urban, editor of the *Gentleman's Magazine,*

> The last transit of the planet Mercury over the sun was so rare a phaenomenon, that 'tis no wonder it excited the curiosity of a considerable number of skilful persons to make due preparations for observing it, by taking care to regulate their time keepers, either by altitudes of the sun or stars, or by meridian instruments, whose exact position had been determined by repeated correspondent altitudes.[10]

The report included an observation by Smeaton.

James Short submitted a report to the Royal Society, which also included an account of Smeaton's observation:

Mr Smeaton in Furnival's-Inn-Court, Holborn, observed the total egress* at 10h 8m 30s, by Mr. Short's clock, thro' a six foot refracting telescope.—He suspects his time some seconds too late, a cloud having just passed off the sun, when he perceived Mercury was gone.[11]

Fig.10 A view of the Earth from Mercury at the start of the 1753 transit (left) and at the finish (right). Since the Earth is rotating eastward, it can be seen that from London Smeaton would only be able to see the later stages of the transit.

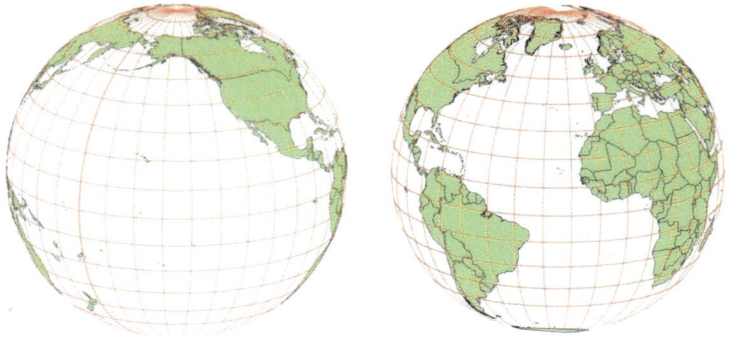

Transits of Mercury across the face of the Sun occur 13 or 14 times a century, so although they are not extremely rare, only ten took place during Smeaton's lifetime and not all of these were visible from the British Isles. The one in 1753 lasted about eight hours, but since the first contact of Mercury's silhouette with the limb of the Sun took place several hours before London sunrise, Smeaton was only able to observe the later stages.

A transit of Mercury is necessarily a telescopic spectacle: It cannot be seen by the unaided naked eye, because the apparent size of the planet is too small—less than $^1/_{150}$ the size of the Sun's disc.

The first time ever that one was seen was in 1631. This was by the French astronomer, Pierre Gassendi. Maybe the most impactful such transit, however, was the one observed by the 20-year old Edmond Halley (1656-

* *Egress*: the departure of the planet's silhouette from the Sun's disk.

1741) from the remote South Atlantic island of St Helena on 7 November 1677. It was the sight of this that gave him a clear idea of how to make an ingenious use of the far rarer transits of *Venus*—next scheduled to take place nearly a century later (in 1761 and 1769): 'This sight', he said, referring to the transit of Venus, 'which is by far the noblest astronomy affords, like the secular games, is denied to mortals for a whole century, by the strict laws of motion. It will be ... shown, that by this observation alone, the distance of the Sun, from the Earth, might be determined with the greatest certainty...'

We will see later what all this would mean for John Smeaton.

The Civil Engineer

S meaton's career now turned decisively to engineering. He put his hand to the design of water-mills, masonry bridges and drainage schemes. His scientific approach to these commissions – at once both analytical and experimental – made its mark with other fellows of the Royal Society. So much so, that when the second lighthouse on the Eddystone Rock was destroyed by a storm in 1755, the President of the Royal Society (Lord Macclesfield) advised the owners that Smeaton – an engineer with no previous experience of lighthouse construction – was the ideal man to take charge of the design and building of a new one!

By October 1759 the new lighthouse – the third on the Eddystone Rock – was permanently operating and ultimately it became the archetypal symbol of civil engineering in Britain, featuring on the crest of the Institution of Civil Engineers.

Around 1760 Smeaton, his wife Ann, and their baby daughter moved back to Austhorpe Lodge in Leeds and set about its renovation.

The demand for Smeaton's engineering skills continued to escalate and inevitably took up most of his time …

BUT astronomical interests were not neglected.

Fig.11 (opposite) The third Eddys- tone Lighthouse (detail of work by Vilhelm Melbye)

In the May of 1768 the Royal Society received two papers from Smeaton on astronomical topics. They were introduced to the Society by the Astronomer Royal, Nevil Maskelyne (1732-1811), who explained that they were from his 'ingenious, and much esteemed friend, Mr. John Smeaton'. He went on to say that the papers would 'prove acceptable presents to astronomers'.

Orbit of the Earth-Moon System

The first paper had a lengthy title, which will seem obscure without further elaboration: namely, *A Discourse concerning the Menstrual Parallax, etc.* Here Smeaton takes as his starting point Newton's explanation that it is not the *centre* of the Earth which orbits the Sun

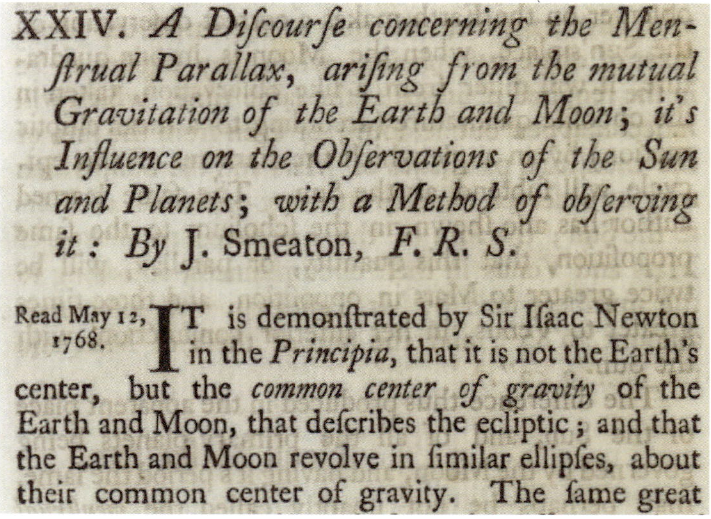

Fig.12 Smeaton's paper on the 'menstrual parallax' (Philosophical Transactions of the Royal Society, v.58 (1768), 156-69)

XXIV. *A Discourse concerning the Menstrual Parallax, arising from the mutual Gravitation of the Earth and Moon; it's Influence on the Observations of the Sun and Planets; with a Method of observing it :* By J. Smeaton, *F. R. S.*

Read May 12, 1768. IT is demonstrated by Sir Isaac Newton in the *Principia*, that it is not the Earth's center, but the *common center of gravity* of the Earth and Moon, that describes the ecliptic ; and that the Earth and Moon revolve in similar ellipses, about their common center of gravity. The same great in an ellipse, but the common centre of gravity of the Earth-Moon *system*.[12] The plane of the orbit is known as the *ecliptic* plane.[13]

Newton's proposition was that the Earth and Moon are orbiting each other: rather like an adult and child skater holding hands and spinning around each other on the ice. To be sure, the Earth, being the larger and more ponderous body, is moving at a slower speed along a shorter path, whereas the Moon (only a fraction of the mass of the Earth) is describing a much longer path at a much greater speed.

In fact both bodies are orbiting around their common centre of gravity, or *barycentre* (B in fig.13) and it is this barycentre that is orbiting the Sun in an ellipse.

Newton thought that point B would be 1.5 Earth radii from the Earth's centre, C. Smeaton, however, using more up to date estimates for the masses of the Earth and Moon, realised that the point B, around which both bodies orbit, would actually be about 0.8 of an Earth radius from C—i.e. within the body of the Earth.

Nevertheless, this position is sufficient to ensure that the direction of the Sun (or any planet), as seen from the

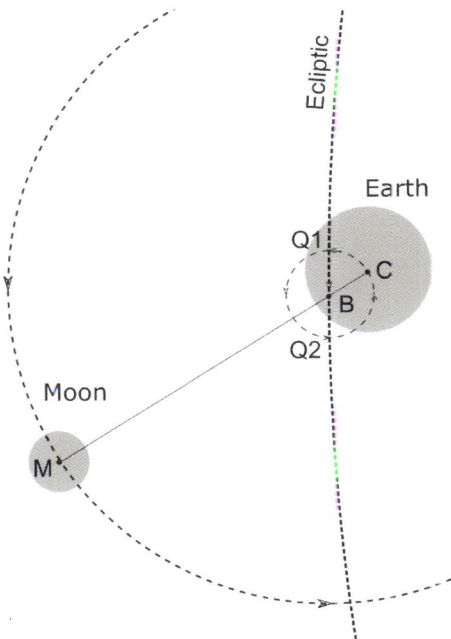

Fig.13 The Earth and the Moon spin around point B, which moves downward along the ecliptic path. The centre of the Earth, C, can be at Q1 (behind B) or at Q2 (ahead of B).

The Sun is in the distance to the right.

centre of the Earth, will be slightly different, depending on the position of the Moon. Clearly, the difference will be greatest when the moon is a waxing or waning quarter-Moon (i.e. at *quadratures*).

Any difference of apparent direction of an object due to changed vantage point is known as the *parallax* and in this case, given the dependence on the position of the Moon and the monthly periodicity involved, it was known as the *menstrual parallax*.[14]

Smeaton notes how this parallax will affect observations of the Sun's position in the sky and in this paper he calculates the quantitative effect and explains how to minimise it.

Out of the Meridian

Smeaton's second paper of 1768 was titled *Description of a new Method of observing the heavenly Bodies out of the Meridian*. It gave his thoughts on how to overcome an observational difficulty.

Fig.14 Smeaton's paper on observing 'out of the meridian' (Philosophical Transactions of the Royal Society, v.58 (1768), 170-73)

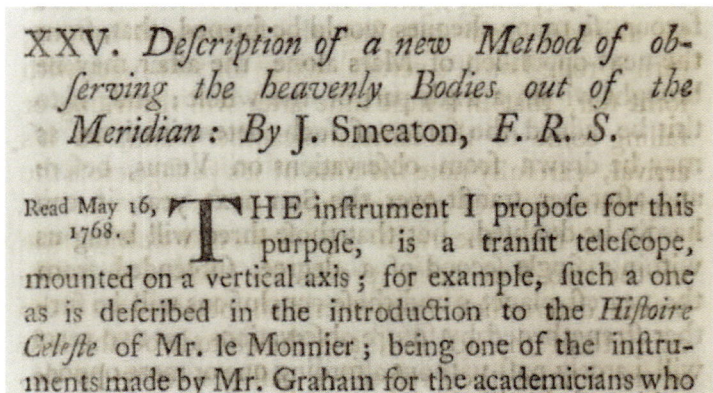

XXV. *Description of a new Method of observing the heavenly Bodies out of the Meridian:* By J. Smeaton, F. R. S.

Read May 16, 1768. THE inftrument I propofe for this purpofe, is a tranfit telefcope, mounted on a vertical axis ; for example, fuch a one as is defcribed in the introduction to the *Hiftoire Celefte* of Mr. le Monnier ; being one of the inftruments made by Mr. Graham for the academicians who

Since antiquity the key pre-occupation of observational astronomers was measuring and recording the position of stars and planets. This was most effectively done by measuring the time and altitude of a celestial body, say a planet, at the moment it was exactly due south of the observer's location and hence at its apparent highest in the sky: In other words, when the body transited the local *meridian*—an imaginary north-south line running across the sky (from the North Celestial Pole to the

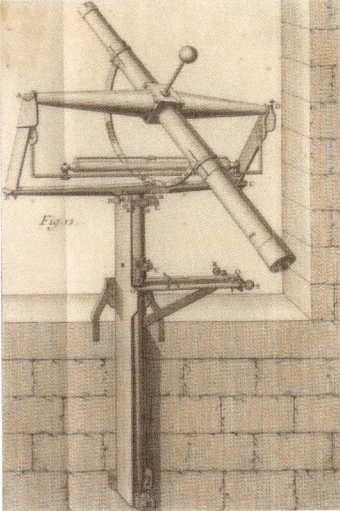

Fig.15 *The type of telescope proposed for the 'out of the meridian' observations (from Le Monnier's* Histoire Céleste *(1741))*

South Celestial Pole). This was usually done with a precision sighting instrument, a quadrant, attached to the face of a north-south running wall or with a telescope mounted in such a way that it could only face south (usually rotating about a fixed axis). A great deal of prior effort would have been invested in marking out the meridian line.

This approach was all very well for the fixed stars at a time of the year when they were well-placed in the night sky. Sometimes, however, observations were needed before (or after) a body had crossed the meridian. For example, planets such as Mercury or Venus never stray far from the Sun in the sky and, therefore, only cross the meridian when the Sun is high in the sky and it is then difficult to see stars and planets. In such cases accurate measurement of the position becomes problematic.

It was this issue that Smeaton chose to address in his paper. He did it by suggesting an observing procedure in which the passage of the body across the sighting hairs in a telescope, or telescope-mounted micrometer, was timed. The time was then compared with the times for a couple of stars of known position that had crossed a few minutes before and afterwards. The chief merit of his method was that it dispensed with the need to rely on the crude setting circles of the telescope to determine the altitude or azimuth—it depended only on a reliable clock.

Partial Solar Eclipse, 1769

On 4th June 1769 Smeaton made careful measurements of a partial solar eclipse from his Austhorpe observatory, using a Dollond refracting telescope (3½ feet focal length) and a micrometer. This wasn't just an idle observation of a curious event: It was a thoroughly scientific, quantitative tracking of the motion of the Moon against the background disk of the Sun, with accurate timings.

Fig.16 Smeaton's report on the 1769 partial eclipse (Philosophical Transactions of the Royal Society, v.59 (1769), 286-88)

XL. *Obfervation of a Solar Eclipfe the 4th of* June, 1769, *at the Obfervatory at* Aufthorpe, *near* Leeds, *in the County of* York. *By* J. Smeaton, *F. R. S.*

	h	′	″
BEGINNING by mean time, A. M.	6	33	1
Middle	7	26	38
End	8	20	16
Total duration	1	47	15
Digits eclipfed		6	46

N. B. The beginning and end of the eclipfe were obferved by an excellent 3½ feet treble object-glafs

In the eighteenth century eclipse timings were an important way to establish the longitude of the places from which observations were made. Such timings were also useful in refining knowledge of the motion of the Moon: knowledge which had an extremely practical bearing for navigation at sea. Although John Harrison and others were developing successful ocean going clocks and watches by the 1770s, these would be prohibitively expensive for decades to come. In the meantime ships' navigators had to rely on the Moon's distance from known stars to provide a 'clock in the sky'. Precise knowledge of where the Moon should be at specific times allowed the time to be deduced from observation. Thus, using the complex calculations of the method of 'lunars', a ship's position could be reliably ascertained.

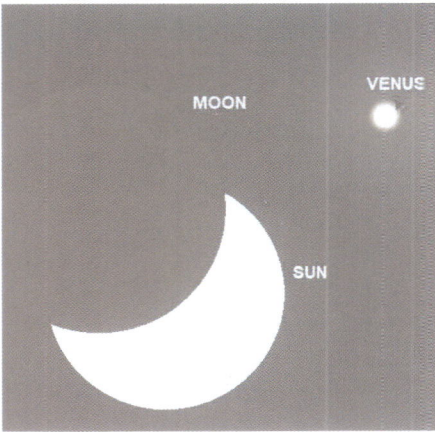

Fig.17 The appearance that the partial eclipse of 1769 would have had, as seen from Austhorpe (magnitude 0.576)

It was in this spirit of precision measurements that Smeaton made his observations of the progress of the partial eclipse that day. How well did he do? In order to answer that question we need to note that there are two critical instants during the eclipse that are susceptible of careful timing. These are the moments when the Moon makes first contact with the limb of the Sun (position P1 in fig.18) and then when the trailing limb of the Moon finally leaves the Sun's disk (position P4 in fig.18).

As can be seen from the beginning of his report in the *Philosophical Transactions* (fig.16), he records these two instants, in "mean time", as:

P1   06 hrs 33 mins 1 secs a.m.
P4   08 hrs 20 mins 16 secs a.m.
(duration 1 hour 47 mins 15 secs)

Precision calculations with modern knowledge of the lunar orbit, deployed by the French *Institute for Celestial Mechanics and Computation of Ephemerides* (IMCCE) give the following values for the eclipse at Austhorpe in 'Universal Time' (which, for practical purposes here, is the same as Greenwich Mean Time, GMT):

P1   06 hrs 38 mins 44 secs a.m.
P4   08 hrs 26 mins 05 secs a.m.
(duration 1 hr 47 mins 21 secs)

At first sight, one might think that Smeaton has erred significantly. This would be wrong, however.

We have to remember that in Smeaton's day there was no standardised time. Mean time (i.e. clock time) was specific to each location. GMT as a civil time standard

Fig.18 The Position of the Moon at key stages during the eclipse: P1 (first contact); M (maximum eclipse); P4 (final contact)

throughout the UK was not introduced until the nineteenth century. Every locality had its own mean time. Because the Earth is spinning on its axis at a rate of 15° per hour and the longitude of Austhorpe is somewhat west of Greenwich, then noon (the moment when the Sun is due south) occurs 5 mins 45 secs later at Austhorpe. Therefore, this is the amount that needs to be added to Smeaton's values to convert his times to GMT, as follows:

P1   06 hrs 38 mins 46 secs a.m.
P4   08 hrs 26 mins 01 secs a.m.
(duration 1 hr 47 mins 15 secs)

Now Smeaton's measurements can be seen as a wonderfully creditable performance - only 2 seconds out at first contact and 4 seconds at final contact!

Transit of Venus, 1769

The Editor of the *Philosophical Transactions* added an interesting incidental note at the end of Smeaton's eclipse report:

> Mr. Smeaton was prevented, by clouds, from observing the entrance of Venus upon the Sun, the evening before.[15]

This was a reference to the much anticipated transit of the planet Venus across the face of the Sun on 3 June 1769 that Halley had talked about (see page 15). This must have been a matter of considerable disappointment for John Smeaton. Unlike transits of Mercury, transits of Venus are very rare: they occur in pairs separated by eight years, but then are not seen again for more than a century.[16] Only the end of the preceding 1761 transit had been visible in England. Timings of it from Wakefield had nevertheless been good enough to determine the difference of longitude between Wakefield and Greenwich. In fact, the resulting longitude of Wakefield was used by Smeaton himself in order to work out the longitude of Austhorpe.

Before 1761 only two people in the world were known to have observed a transit of Venus (that of 1639). These were a twenty-year old from Toxteth, near Liverpool, Jeremiah Horrocks (c.1619-41) and his friend, William Crabtree (1610-44), from Salford. As a result of their observations, the illustrious stronomer, Halley was able to predict where one would have to be in the world to see the next ones (1761 and 1769) in their entirety. Much more importantly, he pointed out that the relative nearness of Venus to the Earth (compared with Mercury) meant that there would be a more noticeable shift in its apparent track across the Sun and the timings, if observed from widely

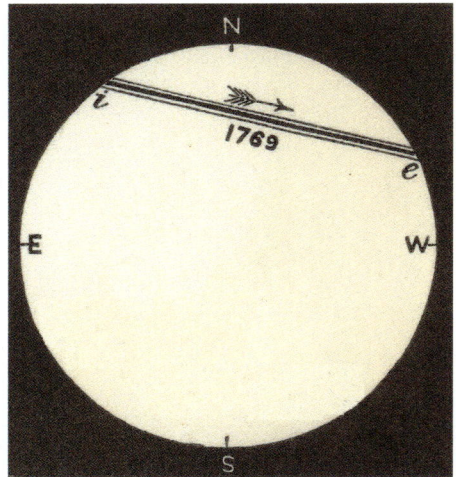

Fig.19 The track of the 1769 transit of Venus across the Sun. The bold central line is as the track would be seen from the centre of the Earth. The upper and lower lines are as seen from from southern and northern hemisphere sites, respectively. The displacement is due to the parallax effect.

separated locations on the Earth's surface. He proposed that this phenomenon be used to triangulate the distance of the Sun and that astronomers of different nations should collaborate in order to achieve this. Halley even calculated exactly where the most propitious far-flung places would be, and in a famous paper of 1716 (written at the age of 60), knowing that he would certainly be dead by the next transit, he declared:

I recommend it, therefore, again and again, to those curious Astronomers, who (when I am dead) will have an opportunity of observing these things, that they would remember this my admonition, and diligently apply themselves with all their might to the making of this observation; and I earnestly wish them all imaginable success; in the first place that they may not, by the unseasonable obscurity of a cloudy sky, be deprived of this most desirable sight; and then, that having ascertained with more exactness the magnitudes of the planetary orbits, it may redound to their eternal fame and glory.[17]

Halley died at a grand old age in 1741, but his exhortation and detailed plans, projected beyond the grave, caused well-equipped expeditions to set out to distant corners of the Earth to observe the 1761 and 1769 transits. The prize, if all went well, would be nothing less that the distance of the Sun—what Halley termed the "solution of a problem the most noble, and at any other time not to be attained to".

For various reasons, the 1761 results were not ideal: Consequently, much rested on the success of the 1769 expeditions.[18]

Given the romance of the transit of Venus story, Smeaton's disappointment at missing the sight of the transit from Austhorpe must have been acute. Nevertheless, he still made

Fig.21 A modern replica of HM Bark Endeavour, in which Captain Cook sailed to Tahiti to observe the 1769 Transit of Venus

an important contribution to the international observing effort, for it was he who had designed the portable observatories for the purpose, taken to Tahiti by Captain Cook and to Hudson's Bay by William Wales and Joseph Dymond.

Fig.20 The mobile observatory tent designed by Smeaton for the 1769 Transit of Venus expeditions to Tahiti and Canada

When the results of all the expeditions were collated, the Oxford astronomer, Prof. Thomas Hornsby (1733-1810), deduced a mean distance for the Sun of 93,726,900 miles.[19] Having found the distance to the Sun, it was also possible to use Kepler's laws of planetary motion to calculate the distance of every other planet. So, it is reasonable to say that Smeaton played a small role in sizing the solar system itself.

Observatory design

The *mobile* observatories for expeditionary use were an unusual departure for Smeaton. In fact, he became an experienced designer of *fixed* observatories for some of the most distinguished astronomers of his day. What is more, they were pioneering designs: Ones which benefitted from his own observing practice.

As we have seen above, the normal activity of astronomers, prior to Smeaton's time, was *positional* astronomy, based on recording the times and altitude of celestial bodies crossing the observer's meridian. For this purpose, all that was required of an observatory was that it should have a clear view of the southern sky and have a telescope that was mounted in such a way as to always be aligned with the local meridian. By the middle of the eighteenth century, however, wealthy amateur astronomers (there were very few professional astronomers) were starting to acquire large telescopes with the aim of examining the physical appearance of objects (moon, planets, nebulae, and comets)—not just the position. They required observatories and telescope mountings that could lend themselves to observation in any direction.

Answering this demand, Smeaton designed some of the earliest observatories in the country fitted with rotating – or rotative – roofs. The rotating domed roofs, with a slot-like opening, which we now think of as the normal appearance of an observatory, were in large part first pioneered by John Smeaton. According to the *Cyclopaedia* of Abraham Rees (1819), "The construction which has been most generally adopted [rotatory roofs]…was contrived by the ingenious Mr. Smeaton, one of whose papers, of the date of 1788, now lying before us was designed for the late Mr. Aubert's observatory at Highbury".[20] This was the observatory whose equipment was stated by William Herschel, the discoverer of Uranus (1781), to be "a treasure beyond value".[21] Besides Aubert's observatory, Smeaton also designed others, such as Matthew Raper's (at Thorley

Fig.22 The York Observatory

Hall) and the moveable roof of Lord Alemoor's observatory at Hawkhill, near Edinburgh, which was also used for a transit of Venus observation in 1769.

After the death of Alexander Aubert, the Rev. Dr William Pearson, of South Kilworth, Leicestershire, purchased (1806) the rotating conical roof that Smeaton had designed for him. It is thought that it was possibly later used for the roof of the York Observatory (fig. 22) constructed in the Museum Gardens by York Philosophical Society (of which Pearson was a member) in 1829.[22]

Astronomical Frontiers

From his house in Austhorpe, John Smeaton maintained regular correspondence with long-term scientific friends. He also paid visits to some, including William Herschel (1738-1822), the professional musician who discovered the planet Uranus, constructed catalogues of the nebulae and made the first attempt to map the distribution of stars in our own galaxy – the Milky Way.

In November 1785, Smeaton visited the Herschels at Clay Hall and was permitted single-handedly to make a sweep of the constellation Eridanus, using the 20-foot Newtonian reflector. The result of the sweep was credited in Caroline Herschel's notebook as "Partly Mr. Smeaton's observation". Use of this telescope was not for the feint-hearted. It involved standing at the top of a ladder about 3.5 metres above the ground to access the eyepiece—a rather dangerous enterprise in the dark.[23]

Amongst his closest friends was the Reverend John Michell (1724-93)—his exact contemporary—of Thornhill, near Dewsbury, about twenty kilometres away. Michell had been the Woodwardian Professor of Geology at Cambridge, but in 1767 had moved to Thornhill to take up a post as rector of the parish church. He was a kindred spirit, fascinated by astronomy. Smeaton visited and stayed with him on occasions. Often the two would seek each other's views on astronomical issues and Smeaton's advice was particularly sought on the building of a great telescope by his Thornhill friend.

In 1767, the year he moved to St Michael's, Michell sent a paper to the Royal Society, which suggested for the first time the existence of true physical binary stars—i.e.

Fig.24 The Parish Church of St Michael and All Angels at Thornhill, near Dewsbury, where Smeaton's friend, John Michell, was the rector

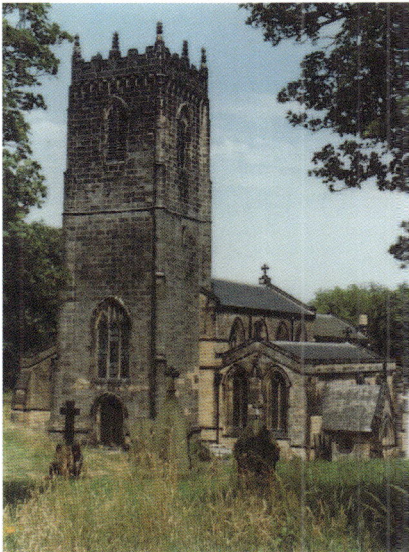

stars orbiting each other. He pointed out that the frequency distribution of the angular separations of known double stars deviated radically from what might be expected if double stars were simply chance 'line of sight' effects amongst stars uniformly distributed in space. According to him

> the natural conclusion from hence is, that it is highly probable, and next to a certainty in general, that such double stars as appear to consist of two or more stars placed very near together, do really consist of stars placed nearly together, and under the influence of some general law … to whatever cause this may be owing, whether to their mutual gravitation, or to some other law or appointment of the Creator.

In 1783 Michell sent a paper to the Society in which he postulated the existence of 'black holes': Stars so massive that, due to gravitational attraction, even starlight would not be able to escape from them. The existence of these entities was not proven until late in the twentieth century.

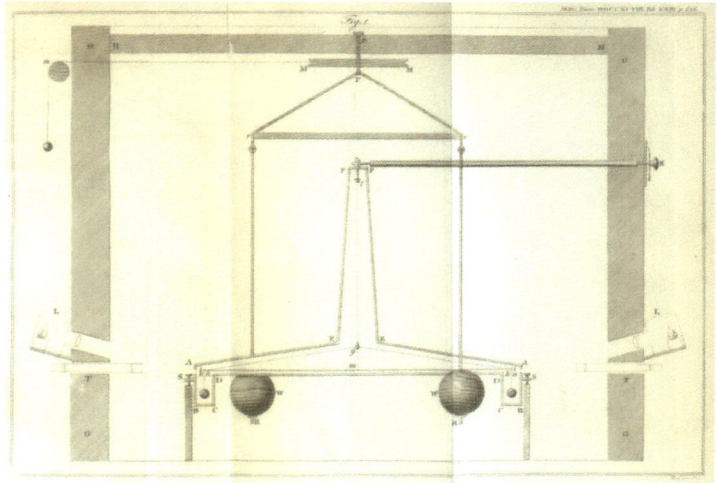

Fig.25 Michell's torsion balance, as modified by Henry Cavendish to 'weigh the Earth'

REVD
JOHN MICHELL BD. FRS
1724 – 1793
GEOLOGIST AND ASTRONOMER
RECTOR OF THORNHILL 1767 – 1793
HE EXPERIMENTED ON MAGNETISM AND
ASTRONOMY, ALSO MAKING A TORSION BALANCE
TO WEIGH THE WORLD. HIS VISITORS HERE
INCLUDED HENRY CAVENDISH,
WILLIAM HERSCHEL, JOSEPH PRIESTLEY
AND JOHN SMEATON.

Fig.26 The Institute of Physics blue plaque at Thornhill parish church, mentioning Smeaton's friendship with Michell

Michell is credited with the first ever realistic estimate of a stellar distance (the star *Vega*, in 1784). He also devised at Thornhill an experimental apparatus—a sophisticated 'torsion balance'—with which he proposed to determine the universal constant of gravitation (G). Given the precision mechanical features of its design, it would have been only natural for him to have discussed its details with his friend Smeaton—erstwhile maker of philosophical instruments and one of the most illustrious engineers in the land. After the death of Michell, Henry Cavendish acquired the torsion balance and, after refining it, used it successfully in a famous experiment to 'weigh the Earth'.[24]

As it happened, Smeaton had already had a peripheral involvement in a previous attempt to 'weigh the Earth'. In 1772 he had supplied a plan of the land around Helvellyn and Skiddaw, in the Lake District, to Nevil Maskelyne for a project inspired by a negative conjecture by Isaac Newton in *A Treatise of the System of the World* (1728).[25] Newton had surmised that the force due to gravity was so weak that everyday terrestrial objects of modest mass would never experience any measurable attraction to each other: "Nay whole mountains", he said "will not be sufficient to produce any sensible effect". By Smeaton's time, however, the precision of scientific instruments had become sufficient to bring Newton's conclusion into question.

Maskelyne's audacious proposal was to measure the gravitational attraction of a hill or mountain, using a plumb-line and careful astronomical observations.[26] In the event, a different mountain was chosen in preference to the ones on Smeaton's map. That mountain was Schiehallion, in Scotland, and the experiment was successfully completed in the summer of 1774.[27]

It can be said, with a fair amount of confidence, that Smeaton, in his conversations with Herschel, Maskelyne and Michell, would surely have been involved in discussing observations and experiments at the very forefront of astronomical thinking.

But discussions ranged far beyond celestial topics too: Michell helped Smeaton to revise the text of his famous book on the Eddystone Lighthouse. Writing reports was never Smeaton's *forte* and he claimed that writing this particular book was harder than building the lighthouse itself. He even asked Michell to recommend a helpful book of 'English Grammar'.[28] What particularly delighted Smeaton, however, were Michell's suggestions for amendments to his account, at the beginning of the book, of the Alexandrian *Pharos*—the famed lighthouse of Antiquity (more than seven times the height of its Eddystone descendant!), which was numbered one of the 'wonders of the Ancient World'. "All your amendments there," he wrote to Michell, "I have adopted, and will make me cutt a figure in Antiquity I didn't expect".[29]

Graduation of Astronomical Instruments

Arguably the most influential paper of John Smeaton on an astronomical theme was his *Observations on the Graduation of Astronomical Instruments* of November 1785. He commenced the essay with the judgement that

> Perhaps no part of the science of mechanics has been cultivated by the ingenious with more assiduity, or more deservedly so, than the art of dividing circles for the purposes of astronomy and navigation.

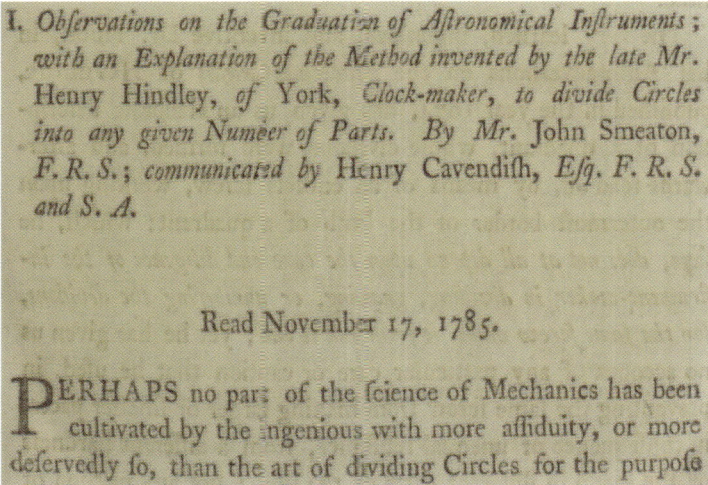

I. *Obfervations on the Graduation of Aftronomical Inftruments; with an Explanation of the Method invented by the late Mr.* Henry Hindley, *of* York, *Clock-maker, to divide Circles into any given Number of Parts. By Mr.* John Smeaton, *F. R. S.; communicated by* Henry Cavendifh, *Efq. F. R. S. and S. A.*

Read November 17, 1785.

PERHAPS no part of the fcience of Mechanics has been cultivated by the ingenious with more affiduity, or more defervedly fo, than the art of dividing Circles for the purpofe

Fig.27 Smeaton's report on the Graduation of Instruments (Philosophical Transactions of the Royal Society, v.76 (1786), 1-47)

He was correct in identifying the crucial role in astronomical discovery that has been (and indeed is) played by the accuracy and precision with which the scales on instruments can be made.

Of course, because astronomers tended to be measuring angles between objects, the graduated scales were naturally engaved on arc-shaped, or circular, scales.

The division of those arcs with precisely engraved equi-spaced marks was by no means a simple exercise. A good deal of investigation had gone into the best ways of achieving this.

That paper, which Smeaton submitted to the Royal Society, was a 47-page masterly historical survey of techniques that had been used from the time of John Flamsteed (1646-1719) and Abraham Sharp (1653-1742) at the Royal Observatory, Greenwich, until the best practitioners of his own day. In particular, he took the opportunity to outline the innovative method of his late friend, Henry Hindley, of York. He had bought Hindley's dividing engine in early 1785 and restored it in his Austhorpe workshop. Since Hindley had been dead some fourteen years, he no longer felt any duty to respect trade secrets embodied in its construction and use.

As an erstwhile maker of precision instruments himself, Smeaton was appreciative of the efforts made by his predecessors in this art and considerate in his praise of their work—with which he sometimes had had first hand acquaintance. For example, referring to his fellow Yorkshireman, Abraham Sharp, he wrote, "I look upon Mr Sharp to have been the first person that cut accurate and delicate divisions upon astronomical instruments…he retired [from Greenwich] to the village of Little Horton, near Bradford, Yorkshire, where I have seen not only a large and very fine collection of mechanical tools, the principal ones being made with his own hands, but also a great variety of scales and instruments made with them, both in wood and brass, the divisions of which are so exquisite as would not discredit the first artists of the present times.[30]

Some time in 1786, Smeaton became aware of an appeal by the French astronomer Jerome Lalande for accurate observations of the planet Mercury when at its greatest angular separation from the Sun – specifically in August and September of that year. Lalande had around that time read to the Royal Academy of Sciences in Paris a fifth Memoir on his theory of the Mercurial orbit, but still needed more data to bring it to completion.[31]

In August, Smeaton tried to oblige. He set up his telescope in August a few days before and after the day when Mercury was expected to attain its maximum *eastern* elongation (around 11th August), intending to time the passage of the planet as it crossed the meridian. He knew this would be a difficult thing to do, since it would be before sunset and the sky would be still bright, but he had seen Mercury on the meridian before, with

XXXIII. *Account of an Observation of the Right Ascension and Declination of Mercury out of the Meridian, near his greatest Elongation*, Sept. 1786, *made by Mr.* John Smeaton, *F. R. S.* *with an* Equatorial Micrometer, *of his own Invention and Workmanship; accompanied with an Investigation of a Method of allowing for* Refraction *in such Kind of Observations; communicated to the Rev.* Nevil Maskelyne, *D. D. F. R. S. and Astronomer Royal, and by him to the Royal Society.*

Read June 7, 1787.

M DE LA LANDE having announced to some of my astronomical friends the utility of accurate observations of Mercury, at his two elongations the last year, in August and

Fig.28 Smeaton's report on the Greatest Elongation of Mercury (Philosophical Transactions of the Royal Society, v.77 (1787), 318-43)

the self-same telescope—one that he had made himself in 1768. This time, however, he had no luck and resolved to try again in September, when the maximum *western* elongation would occur and Mercury would rise in the east before the Sun.

On 23rd September 1786, the day in question, the air was clear and serene. He started observing at 5.15 a.m, three-quarters of an hour before sunrise. Although the sky was already quite bright, Mercury was easily found with an opera glass and presented a clean crisp image in his telescope. Using a micrometer, he began a process similar to that he'd employed in 1768 for observations of heavenly bodies 'out of the meridian'. This time, however, his telescope mount was such that, once Mercury was nicely placed within the (1° 17′) field of view and its passage across the wires of the micrometer had been exquisitely timed, he found he could leave the telescope until the evening without its orientation being in the slightest disturbed. Thus, in the evening, he was able to time the passage stars of known-position (λ *Ceti* and *o Tauri*) across the same wires. In this way, by applying the known rotation rate of the Earth, he was able to calculate what had been the true position of Mercury at the time of his morning observation. In fact, the stability of the telescope mount was so firm that when he re-examined the orientation on 30th September it had not budged. He then had to set off on a lengthy journey and locked the door of the Austhorpe observatory. Returning on the 13th October, he was astonished to see that the telescope direction had suffered no more apparent alteration!

In his *Introduction to Practical Astronomy* (1829), William Pearson (1767-1847) gave a lengthy account of

Although Smeaton's method of finding the accurate position of Mercury sounds straightforward in principal, the method of taking the readings from the micrometer and clocks was very involved. It took more than 40 pages of his report to explain the procedure and present the calculations to derive the position of Mercury. Along the way, he also gave a careful analysis to determine what corrections were needed to allow for refraction (or bending) of the light as it passed through the Earth's atmosphere. Mercury had only been risen an hour or so, and was therefore quite low in the sky—a circumstance which guarantees that refraction will be significant.

The Quadrant of Altitude

His last sustained piece of writing on an astronomical matter was a report describing a patented improvement (a 'quadrant of altitude') that he had devised for the celestial globe and how it could be used to solve problems in positional astronomy.

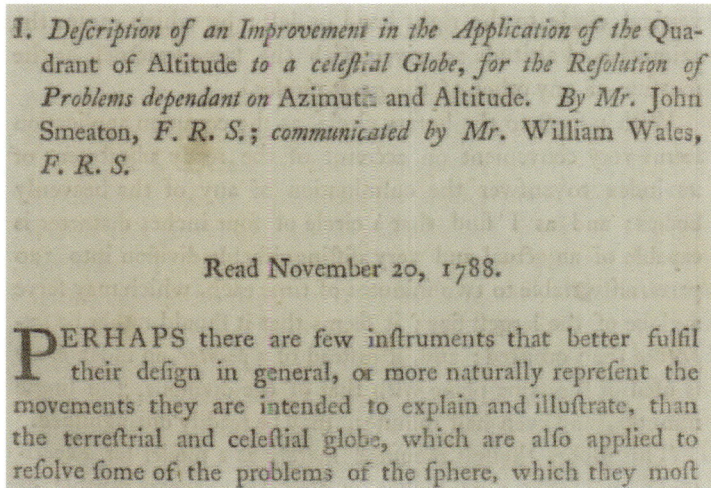

I. *Defcription of an Improvement in the Application of the* Quadrant of Altitude *to a celeftial Globe, for the Refolution of Problems dependant on* Azimuth and Altitude. *By Mr.* John Smeaton, *F. R. S.; communicated by Mr.* William Wales, *F. R. S.*

Read November 20, 1788.

PERHAPS there are few inftruments that better fulfil their defign in general, or more naturally reprefent the movements they are intended to explain and illuftrate, than the terreftrial and celeftial globe, which are alfo applied to refolve fome of the problems of the fphere, which they moft

Fig.29 Smeaton's Quadrant of Altitude (Philosophical Transactions of the Royal Society, v.79 (1789), 1-6)

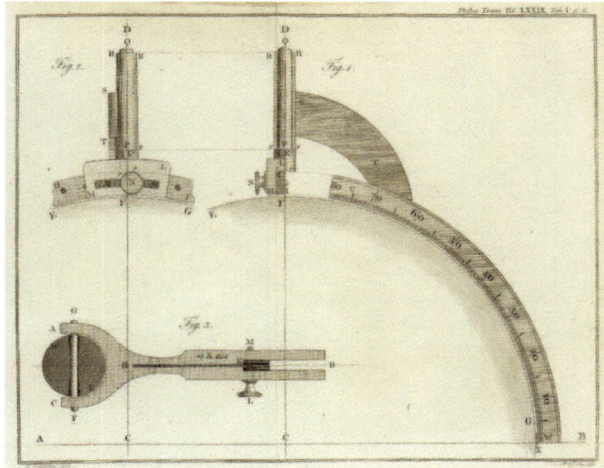

Fig.30 The Quadrant of Altitude fitted to a globe

Nothing more on this emerged from the pen of the Austhorpe astronomer, for around the year 1785, his health had started to decline noticeably and he had resolved to withdraw from any formal business as much as he could. By late 1791 he was in almost total retirement and for some while had been devoting himself solely to writing.

John Smeaton died on 28 October 1792, six weeks after having a stroke as he walked through his beloved garden at Austhorpe. According to his daughter, "He always apprehended the stroke, as it was hereditary in his family; he dreaded it only as it gave the melancholy possibility of out-living his faculties, or the power of doing good: to use his own words, 'lingering over the dregs, after the spirit had evaporated!'".

Thankfully, however, his memory and intellect during his final weeks seem to have been spared. His daughter described the end in moving words tinged with astronomical allusions: "He would sometimes complain of his own slowness (as he called it) of apprehension, and then would excuse it with a smile, saying, 'It could not be otherwise, the shadow must lengthen , as the sun went down!' There was no slowness in fact to lament; for he was as ready at calculations, and as perspicuous in explanation, as at any former period….. The body gradually sunk, but the mind shone to the last."

Library and Instrument Sale

Not long after Smeaton's death, the London bookseller, John Egerton, announced a sale of book collections, '*including the Library of John Smeaton, Esq., FRS*'. The sale catalogue gave a list of book titles, but unfortunately did not itemise which were from Smeaton. Nevertheless, using various clues, subsequent biographers have been able to deduce 50, or so, of these. They include a fair number of works with an astronomical bearing or on optics: books by James Ferguson, John Flamsteed, David Gregory, Charles Le Monnier, Nevil Maskelyne, and Isaac Newton.

More testimony to the seriousness of Smeaton's astronomical pursuits was the sale catalogue, issued in April 1793, for *the Valuable Collection of Curious Astronomical, Philosophical, Optical and Mathematical Instruments…the truly genuine property of John Smeaton, Esq., Civil Engineer and FRS deceased*. These included at least three telescopes, an equatorial mount, a telescope micrometer (made by Smeaton himself), two transit instruments, a Hindley repeater clock, a celestial globe and a sextant.

A telescope and micrometer made by Smeaton (c.1770) were also given, by his daughter, Mary Dixon, to the mathematician, Mary Somerville. The latter donated these to the Royal Astronomical Society in

42   1845. They are now held by the Science Museum London.[33]

Conclusion

Without doubt, John Smeaton was a capable astronomer. He followed avidly the developments in astronomy, both in Britain and Europe and made modest contributions, both through his various papers— submitted mainly to the Royal Society—and his instrument construction and pioneering observatory design.

It is true that he made no notable discoveries, but his commitment was genuine and long-lasting. His civil engineering achievements were considerable and naturally eclipsed what might be considered his private pastimes. Perhaps this is why very little, if anything, of substance has been written about this side of his life.

Hopefully, this modest booklet will have uncovered for you a little-known aspect of John Smeaton's life and passions: One, moreover, that was very special to him throughout his life.

David Sellers
Figs. 1, 2, 3, 4, 6, 10, 13, 17, 19, 21, 22, 24, 26

Lives of the Engineers by Samuel Smiles (1862)
Fig. 5

The Century Dictionary (1895), p.5240
Fig. 7

The Royal Society Philosophical Transactions
Figs. 8, 9, 12, 14, 16, 25,

Wikimedia Commons
Fig. 11

Histoire Céleste, Charles L Monnier (1741)
Fig. 15

IMCCE
Fig. 18

Wellcome Collection, Engraving by Bernard
Fig. 20

Ivor Trueman
Front cover

[1] James Ferguson, *Astronomy explained upon Sir Isaac Newton's principles and made easy to those who have not studied Mathematics* (1778), 184

[2] Mary Dixon, *Mr John Smeaton*, The European Magazine, and London Review, v.34 (1798), 310

[3] Preface to the 1837 edition of Smeaton's 'Reports' (*Some Account of the Life, Character and Works of Mr John Smeaton, FRS*)

[4] Annual Register, 1793 (published in 1810). *Account of Mr. John Smeaton*, by Mr. John Holmes, Watchmaker, of the Strand, 9

[5] J.R.M. Setchell, *The Friendship of John Smeaton, FRS, with Henry Hindley*, Notes & Records of the Royal Society, v.25 (1970). 79-86

[6] A.W.Skempton (ed), *John Smeaton, FRS* (1981), 9

[7] Benjamin Wilson, *A Series of Experiments on the Subject of Phosphori, etc* (1776), 92

[8] Jérôme Lalande, *Journal d'un Voyage en Angleterre, 1763*, Studies on Voltaire and the Eighteenth Century, v.84 (1980), 72 (tr. DS)

[9] Dr Knight & John Smeaton, *Some Improvements of the Marine Compass*, Philosophical Transactions, v.46 (1749-50), 513-17

[10] *The Gentleman's Magazine and Historical Chronicle*, v.23 (May 1753), 211

[11] James Short, *Observations of the transit of Mercury over the sun, May 6, 1753*, Philosophical Transactions, v.48 (Dec 1753), 199

[12] Isaac Newton, *Principia*, Book 3, propsition 12, theorem 12. Newton explains the principle in relation to the orbit of Jupiter around the Sun.

[13] The plane of the Earth's orbit is known as the *ecliptic plane*, because it is when the Moon crosses this plane that there is potential for it to get in the way of our sight of the Sun and thus create a solar eclipse.

[14] The *menstrual parallax* is now more commonly referred to as the barycentric parallax. The periodicity is approximately 29.5 days - i.e. the synodic period of the Moon (the duration of its orbit around the Earth relative to the position of the Sun). This parallax only needs to be taken into account when performing extremely precise positional calculations.

[15] *Philosophical Transactions of the Royal Society*, v.59 (1769), 288

[16] Subsequent to the one in 1639, there were transits of Venus in the

years: 1761, 1769, 1874, 1882, 2004 and 2012. The next one will be in 2117.

[17] Edmund Halley, *A new Method of determining the Parallax of the Sun, or his Distance from the Earth*, translation given by James Ferguson in *Astronomy Explained, etc*, 1716, 457-458

[18] For a fuller account of the history of the transit of Venus and its significance, see *The Transit of Venus: The Quest to Find the True Distance of the Sun*, David Sellers (2001)

[19] Thomas Hornsby, *The Quantity of the Sun's Parallax as deduced from the Observations of the Transit of Venus, on June 3, 1769*, Philosophical Transactions of the Royal Society, Vol 61, 1771, 579

[20] Rees's *Cyclopaedia* (1819), v.30, 492

[21] Letter from William Herschel to Alexander Aubert, 9 Jan 1782, *The Herschel Chronicle*, by Constance A/. Lubbock, 103

[22] Mike Frost, *Reverend Doctor William Pearson in South Kilworth Leicestershire*, The Antiquarian Astronomer, v.3, 2006, 51-52

[23] Wolfgang Steinicke, V*isitors to the Herschels between 1777 and 1822*, Journal of Astronomical History and Heritage (2023), v.26, 563; and Michael Hoskin, *Discoverers of the Universe* (2011), 92 and 95

[24] Henry Cavendish, *Experiments to Determine the Density of the Earth*, Philosophical Transactions, v.88 (1798), 469-526

[25] Isaac Newton, *A Treatise of the System of the World* (1728), 41

[26] Nevil Maskelyne, *A Proposal for Measuring the Attraction of Some Hill in this Kingdom by Astronomical Observations* (1772), printed in 1776.

[27] Nevil Maskelyne, *An Account of Observations made on the Mountain Schiehallion for finding its Attraction* (1775), printed in 1776.

[28] Letter from Smeaton to Michell, 23 Nov 1785, *Weighing the World: The Reverend John Michell of Thornhill*, by Russell McCormmach, 405

[29] *Ibid.*, 403

[30] William Cudworth, *Life and Correspondence of Abraham Sharp* (1889), 167-68

[31] Jerome Lalande, *Sur la Théorie de Mercure: Cinquième Memoire, Où l'on rectifie les principaux élémens de Mercure, par de nouvelles Observations*, Mémoires de l'Académie Royale (1786), 272-313

[32] William Pearson, *Smeaton's Method of using a wire-micrometer*

with an equatorial stand, Introduction to Practical Astronomy, v.2 (1829), 160-165

[33] Object number 1931-347. 'A refracting telescope of 1½-inch aperture and 36-inch focal length by John Smeaton on an adjustable equatorial mounting with a tripod stand. The telescope is fitted with an integral filar micrometer [RAS No.7]'

APPENDIX

Papers of John Smeaton on astronomical topics published in the Philosophical Transactions of the Royal Society of London.

1. *A discourse concerning the menstrual parallax, arising from the mutual gravitation of the Earth and Moon; it's influence on the observations of the Sun and Planets; with a method of observing it*, v.58 (**1768**), 156-169

2. *Description of a new method of observing the heavenly bodies out of the meridian*, v.58 (**1768**), 170-173

3. *Observation of a solar eclipse the 4th of June, 1769, at the observatory at Austhorpe, near Leeds, in the county of York*, v.59 (**1769**), 286-288

4. *Observations on the graduation of astronomical instruments; with an explanation of the method invented by the late Mr. Henry Hindley, of York, clock-maker, to divide circles into any given number of parts* [with Henry Hindley], v.76 (**1786**), 1-47

5. *Account of an observation of the right ascension and declination of Mercury out of the Meridian, near his greatest elongation, Sept. 1786, made by Mr. John Smeaton, F. R. S. with an equatorial micrometer, of his own invention and workmanship; accompanied with an investigation of a method of allowing for refraction in such kind of observations; communicated to the Rev. Nevil Maskelyne, D. D. F. R. S. and Astronomer Royal, and by him to the Royal Society*, v.77 (**1787**), 318-343

6. Description of an improvement in the applications of the quadrant of altitude to a celestial globe, for the resolution of problems dependant on azimuth and altitude, v.79 (**1789**), 1-6